Fujifilm X-T3: A Complete Guide from Beginner To Advanced Level

Table of Contents

Introduction

In the form of the X-T3, FUJIFILM has created a powerful and versatile mirrorless camera that can be used by photographers of all skill levels. The APS-C X-Trans CMOS 4 sensor, which it uses, has 26.1 megapixels and allows for the shooting of high-resolution photographs with low noise and a broad dynamic range. The camera's focusing technology is powerful enough to keep up with fast-moving objects and other sophisticated capabilities, such as eyes recognition and 4K live video.

A photographer looking for a high-performance camcorder at a reasonable budget will find the FUJIFILM X-T3 to be an excellent option due to its small design and intuitive UI.

Various Camera Components

Mirrorless interchangeable lens camera featuring 4K video recording, an APS-C sized sensor, and a vast ecosystem of lenses and accessories, the FUJIFILM X-T3. The many components of a camera and their relative roles will be discussed in this chapter.

In a first step, we'll check over the camera's exterior, lens mount, viewfinder, LCD screen, shutter click, flash hot shoe, and battery compartment. We'll also go over some of the add-ons available for the FUJIFILM X-T3 that may boost your photographic abilities even further.

The FUJIFILM X-T3 boasts a robust body and an electronic viewfinder as standard features. It also has a place to store batteries, an SD card slot, a display, and a pop-up flash.

These enhancements mean that you may start taking pictures of a professional quality right now.

Body: \s The front and the back are the two basic components of the camera's body. On the camera's front panel, you'll discover a viewfinder magnified by 200% as well as controls for a number of useful features. The power button, shutter release, and four-way controller are all placed here.

The camera's SD card slot, viewfinder cover, and battery compartment (which can hold two rechargeable NiMH batteries) are all accessible from the back.

The Electronic Rangefinder:

You can see what's going on outside the camera's viewfinder on the LCD screen at this spot. It also helps maintain focus inside the camera's lens.

Integrated pop-up Flash:

This is an essential item for ensuring well-lit photos are taken in dim conditions.

Functions to know.

The FUJIFILM X-T3 has several features that make photography easier, such as a better focusing mechanism, 4K video recording, and a tilting LCD screen. What's more, it offers a wide range of shooting modes, from Auto (suitable for beginners) to Manual (preferred by specialists). With to its user-friendly interface and superior picture processing, the FUJIFILM X-T3 is a fantastic choice for any photographer.

Dial mode

The mode dial on a camera is an essential feature since it allows you to choose between different shooting modes. The camera's mode dial helps you easily choose between several shooting modes including Program AE, Aperture Priority AE, Shutter Priority AE, Manual Exposure, Scene Modes, and Advanced Filters. Each mode offers its own selection of customization choices for your images.

On the back of the camera is a button labeled "Fn," which allows you to adjust a variety of additional features. The ISO dial, the white balance switch, and so on are all instances of such settings and dials. The FUJIFILM X-T3 is packed with features and capabilities that will help you take amazing images.

The FUJIFILM X-T3 may be used to a lot of uses, and it comes with a number of features that make it easy to shoot great photographs. You may switch between different shooting modes by twisting the mode dial, and you can access the camera's most important settings by pressing the "Fn" button. With the FUJIFILM X-T3 and all of its features and capabilities, you can take stunning images.

Get crisp images of faraway objects and landscapes, whether rain or sun, with the FUJIFILM X-T3 Digital Camera's powerful optical zoom lens that can magnify by a maximum of 6x.

The F2.8 aperture of the X-fast S10 allows for stunning shallow depth of focus effects. Captivating images of moving subjects or those in poor light are within reach thanks to the camera's continuous shooting capability of 15 frames per second.

Focusing

Multiple options are available on the FUJIFILM X-T3 to help you take photographs of exceptional quality. The focus lever, which enables you to change the lens's focal length, is a crucial component. The focus lever allows you to easily change from auto to manual focusing, giving you more creative leeway in your photographs.

For finer compositional control, you may alter the size and form of the focal point with the focus lever. Additionally, it facilitates subject tracking, making it an ideal tool for action photography.

Dials for command

Since you can adjust the dials without taking your eye off the viewfinder, the Command Dials offer a time-saving convenience. Changing the shutter speed, ISO, and aperture of the camera is as simple as turning a dial. This will make it much simpler to capture any moment in time with excellent clarity and composition. The FUJIFILM X-T3 is a great option for low-light photography because to its high-quality optical viewfinder.

This might help you see the ultimate product of your photography before you take the image. Among the many benefits of this is the improved chance of capturing a well-exposed shot. You may switch between aperture priority and manual mode using the physical buttons on the camera. Anybody looking to take high-quality pictures on the go will find the FUJIFILM X-T3 to be an excellent choice.

Dials for function

The knobs that control the camera's settings are a great addition. These dials provide quick access to all of the camera's most crucial settings, including exposure, shutter speed, ISO, and white balance. Customers may simply spin these dials to get their ideal configuration without having to delve into menus or fiddle with complex choices.

The knobs are labeled so that users may quickly and easily find the optimal settings for any given shooting environment, and then take the finest possible shot. The X-Series cameras are compact and lightweight so the user may take them wherever and use them whenever they choose. The X-compact S10 is perfect for on-the-go photography because to its great picture quality and large image sensor.

This camera is ideal for those who wish to take photographs that are comparable to those taken by professionals but who don't have the time to spend fiddling with the camera's settings and menus. The X-T3 is a portable, lightweight, and quick camera.

All of the X-previous T3's capabilities have been improved upon.

As the Phase AF technique is so precise, focusing takes just 0.03 to 0.5 seconds.

This camera's second-generation high-speed contrast detection autofocus allows for rapid focusing in a wide variety of lighting conditions—up to 16 times faster than conventional autofocus systems.

Lamp warnings

When it comes to cameras, the FUJIFILM X-T3 is hard to beat because to its many useful features. The camera's status may be shown through the Indicator Lamp, which is another useful function. When the camera is ready to take a photo, a light will come on, and it will go dark when it is not. The light may also be used to show that the camera is in standby or to alert users to a problem.

Thanks to this function, photographers may easily remember their preferred shooting parameters and consistently produce high-quality photographs. The picture sensor of the FUJIFILM X-T3 has 2.36 million pixels. This allows the camera to take high-quality pictures regardless of the lighting conditions. In addition, the camera includes an electronic viewfinder, which may make framing shots considerably simpler.

Screen usage

The LCD screen is one of the best parts since it lets you analyze your images and make changes as you go. The LCD screen allows you to see what your photographs will look like, so you can fix things like exposure and composition before you ever snap them. Live preview on the LCD panel allows you to check your framing before you release the shutter.

In terms of picture quality and color accuracy, the FUJIFILM X-T3 stands out as a top-tier LCD screen. It has a 10.2 inch screen with full HD quality (1920 x 1080), making it great for viewing movies or working on video projects. The monitor's stand may be adjusted to your preferred viewing height and angle. The Fujifilm X-flexible S10's color options and brightness make it a great choice for any professional scenario.

Focusing with viewfinder

The FUJIFILM X-T3 has an excellent viewfinder focusing mechanism and is a very capable camera. It gives the user a quick and precise method of zeroing in on the target. The camera has a sophisticated focusing technology that can find and follow moving things in real time, enabling you to get clear shots no matter how dim or dark the lighting is.

Viewfinder features like as focus peaking and magnification make it simple to compose the ideal photo.

The LCD display may be used in either bright sunlight or artificial lighting with no problems in legibility.

The viewfinder

Focus peaking and magnification using a dedicated lens mounted directly to the camera's sensor Rugged, waterproof construction with anti-fog coating

It has a 16.0 x 24.0mm digital sensor and a Penta-core processor.

The Deluxe Underwater Action Camera supports the following memory card types: SDHC/SDXC UHS-I/U3 Class 10, UHS Speed Class 3.

The following are specifications for the 960P Ultra HD with 2x Video Resolution that make it a good option for you to consider.

Monitor use.

The viewfinder choices on the FUJIFILM X-T3 are extensive, making it a powerful and aesthetically pleasing camera. The camera's 3.2-inch LCD screen is bright and crisp, and the touchscreen interface makes it easy to review and edit your photos.

The X-T3 has a high-resolution electronic viewfinder to assist in framing your shots. Under every lighting condition, its refined focusing mechanism will provide sharp images. Whether you're shooting landscapes or portraits, the FUJIFILM X-viewfinder T3's will wow.

Even in direct sunlight, the FUJIFILM X-viewfinder T3's displays an exceptionally clear and brilliant picture of your subject. The electronic viewfinder (EVF) has 2.36 million dots and 0.62x magnification. The EVF has a diopter adjustment to help you focus on your subject, and when you get your eye close to it, the camera switches from the rear LCD to the EVF thanks to an eye sensor. While using the electronic viewfinder (EVF), you may be certain that the image you see is the image you will obtain.

With its excellent resolution and great specifications, the FUJIFILM X-T3 LCD screen is a worthy addition to any digital photographer's arsenal. A whole 16 million colors will be shown on the 3.2-inch LCD display, which has a resolution of 1920x1080.

The display is HDR-compatible, which means it can display a wider range of tones and darker blacks. It has a number of connectors, including HDMI, USB Type C, and DisplayPort, so it can be linked to a variety of peripherals. The X-T3 is your best option if you want a low-cost LCD screen that doesn't sacrifice on quality.

Brightness adjustment

The FUJIFILM X-LCD T3's screen has a simple and rapid brightness adjustment. Hence, you may adjust the screen of the camera to work with the available light, whether you're taking pictures during the day or at night.

The method for doing so is straightforward and simple to grasp. You may adjust these parameters on your FUJIFILM X-T3 to enhance the readability of the Display and the quality of your photographs.

When you push the LCD/EVF button on the back of your camera for a while, the words "LCD/EVF" will show in white on a black background.

After the indicator light goes back to white, you may release your finger.

To restart the camera, press and hold the "Power" button for several seconds. till the cursor turns into a lighted screen. The FUJIFILM X-LCD T3's screen's brightness may be changed in a flash.

Therefore, whether you're filming during the day or the night, you can adjust the screen to the appropriate brightness.

Display rotation.

The screen of the FUJIFILM X-T3 can be rotated 180 degrees, making it ideal for taking self-portraits and selfies. Thanks to this function, you may shoot images from both low and high angles with ease. The fact that the screen can be flipped between landscape and portrait modes makes it useful in any photographic setting. The FUJIFILM X-T3 is a great camera for photographers of all skill levels because to its various capabilities, including a screen that can be rotated.

Modes of display

The display mode is only one of the many ways in which the FUJIFILM X-T3 stands apart from the crowd. You may compare two photographs side by side or zoom in on a specific area with the help of the display mode. Histograms and other shooting data are also included for your convenience. Reviewing and adjusting your shots is a breeze with the Fujifilm X-display S10's mode.

Using the Menus

Access to the camera's settings is simple through the on-screen interface. This section will teach you the ins and outs of the FUJIFILM X-settings T3's menu so you can customize your camera to your liking. Here, you'll learn the fundamentals of using Lightroom, including how to set preferences, activate features like Focus Peaking and Face Detection, and navigate the interface. By reading this guide, you will be able to get the most out of your FUJIFILM X-T3.

One Next Step: Navigating the Menus

The FUJIFILM X-menus T3's are well-organized and straightforward. Most of the camera's controls may be accessed in a flash from either the Main Menu or the Custom Menu.

It's important to note that these windows don't provide you access to all setup choices. Common settings like White Balance and

ISO, as well as some useful capabilities like Focus Peaking or Face Detection, are accessible via a three-step process involving selecting the mode menu option on the FUJIFILM X-T3, then selecting an individual mode on that menu, such as 'AF-C,' and making this process the specific setting you want.

Instead, most cameras' menus have you choose "AF-C" before you can settle on a specific setting (like "EV + 1.0") in the submenu.

Second, verify that your camera's settings are correct.

It might take some time to go through the Fujifilm X-extensive S10's menus and find the setting you're looking for if you're not already familiar with the camera's features and settings. Changing certain settings requires going via the mode menu (described in Step 1), while changing others requires going through a different menu and then selecting from a submenu inside the mode menu (described in Step 2).

You may use the viewfinder on your camera to assist in this procedure. The settings cogwheel is located at the top left of this window and may be accessed by clicking on it. To get to them, use the up and down arrows to navigate the main menu's options, and then hit the right arrow button once or twice to open the submenu for a certain menu item.

Then, choose that option using the directional pad's down and left buttons. You may learn more about the different shooting modes and other camera capabilities by reading the manual or quick start guide you should have downloaded in step 2.

Touch Screen Interaction

FUJIFILM X-T3 is a mirrorless digital camera with touch-screen controls. The camera's three-inch LCD screen transforms into a touchscreen in this mode, facilitating quick and straightforward access to the camera's many customization options.

With the touch screen mode, users may make changes and take pictures without touching any buttons. The touchscreen also allows for quick adjustments to the camera's settings, such as focusing and exposure. Because of its intuitive touch screen, the FUJIFILM X-T3 is a fantastic camera for photographers of all experience levels.

Starting to use your camera.

The FUJIFILM X-T3 is a professional-grade camera capable of capturing stunning moving and static imagery. There are many features in this application, and in order to get the most of them, you need learn the basics. Learn the fundamentals of using your FUJIFILM X-T3 to its full potential.

We'll go through basic camera maintenance, including mounting, adjusting, and experimenting with your camera's settings. You'll be an expert FUJIFILM X-T3 photographer after reading this.

Attaching of strap

The FUJIFILM X-T3 is a top-tier camera that excels in both amateur and professional settings. You may customize it to your needs with a variety of add-ons and adjustments, like this strap. This strap may help you keep your camera close and secure so you can focus on getting the finest picture possible.

Strapping on your FUJIFILM X-T3 is quick and easy. The length of the strap may be quickly changed to provide a comfortable fit around the wearer's neck or shoulder. This accessory will ensure that your FUJIFILM X-T3 is never out of reach. This strap adjusts conveniently and securely.

You can connect your FUJIFILM X-T3 to your camera or a tripod with the strap's two metal attachments, so it's always within reach.

The camera will be safe from the elements whether you're shooting inside or out. Using the fitting that attaches to the bottom of the camera, the strap may be attached or detached in a flash whenever a tripod is involved.

Attachment of lens

The FUJIFILM X-T3 is a mirrorless camera that supports a wide variety of lenses. To maximize the effectiveness of your photographic endeavors, it is possible to choose from a broad variety of lenses. Whether shooting landscapes or wildlife, the

FUJIFILM X-T3 has you covered with its wide-angle and telephoto zoom lenses.

Its camera's portable form factor and intuitive UI make it ideal for both novice and experienced photographers. The variety of lenses for the Fujifilm X-many S10 makes it a fantastic choice as a primary camera for any endeavor. Stunning 12.2-megapixel still shots are possible with the FUJIFILM X-T3, thanks to the camera's new EXR sensor, and the camera's ability to capture high-definition video is a major plus.

Using cards

The FUJIFILM X-T3 is a state-of-the-art camera with a ton of cool extras. If you want to get the most out of your camera, check to see that the battery and memory card are correctly placed. Follow the guidelines in this handbook to insert a battery and a memory card into your FUJIFILM X-T3 camera. This guide will help you get the most out of your camera so you can capture all of life's unforgettable events.

After inserting the battery and memory card, the camera must be switched on and given time to power up. Make sure that the two halves are correctly connected.

Establish the genre of shooting you'll be undertaking. There are four modes to choose from: Manual, Aperture Priority, Shutter Priority, and Program. The manual mode lets you control every aspect of the camera, from the ISO to the exposure.

Live View may be activated by pressing the OK button; once it does, make sure the magnification level is set to 100%. The dial on a camera gives you access to several shooting options (shown).

The camera's dials and buttons may be used for fine-tuning what Program Mode already allows you to do.

Use a tripod, but make sure it's in a spot where it won't get in the way while still supporting your camera.

In the absence of a tripod, place your camera on a table or counter high enough that you won't be able to reach the floor.

Have the option to become low or high by lying down if required.

Battery charging

If you're a photographer in need of a strong and dependable camera, go no further than the FUJIFILM X-T3. The battery life is fantastic, but it's vital to charge it properly to get the most out of it. Here, you'll learn how to get the most out of your FUJIFILM X-T3 battery by charging it correctly.

Step one in charging the battery for your FUJIFILM X-T3 is to plug in the charger.

To get your camera's orange blinking light to appear, press and hold the power button.

The power pack is now being charged.

When the battery is fully charged, your camera's LCD screen will turn green.

You should let your FUJIFILM X-T3 charge for around an hour before picking it back up to use.

Switching On/Off for camera

Knowing how to switch the camera on and off is a crucial skill for utilizing it. In this section, you'll learn the basics of using your FUJIFILM X-T3, including how to switch it on and off and how to make the most of its functions. Reading this chapter will provide you the knowledge you need to use your FUJIFILM X-T3 with ease and confidence.

Activating the FUJIFILM X-T3 camera

To activate the FUJIFILM X-T3, just click the power button on its top.

When powered, the button to switch it on will glow.

Holding the power button for two seconds and letting go will either turn the camera off or restart it. the trigger.

Levels of battery

The FUJIFILM X-battery T3's life is exceptional, enabling you to take up to 800 shots before needing to be recharged. While using the X-T3, you may monitor your battery life on the LCD screen and charge spare batteries using the external charger. Plus, it is powered by USB, so you can charge your camera even while you're on the go.

For photographers seeking for a high-performance camera with a long battery life, the FUJIFILM X-T3 is a fantastic alternative.

Movie filming

The FUJIFILM X-T3 is a great option for filmmakers wishing to capture moving images. Capturing 4K video at 30/25 fps in 10 bits of color depth with 4:2:2 chroma subsampling is possible because to the camera's powerful image processor. Features like face/eye recognition, focus peaking, and manual focus assist make it much simpler to capture high-quality video.

The FUJIFILM X-T3 is a great camera to have if you want to make movies. Capturing 4K video at 30/25 fps in 10 bits of color depth with 4:2:2 chroma subsampling is possible because to the camera's powerful image processor. Features like face/eye recognition, focus peaking, and manual focus assist make it less of a chore to capture high-quality video.

Movie filming

There is no better video camera on the market than the FUJIFILM X-T3. It features a high-quality sensor, records 4K video, and has built-in picture stabilization, making it a great choice for filmmakers.

The camera's user interface is also quite intuitive, making it a breeze to operate. The FUJIFILM X-advanced T3's image processing technology and sturdy focusing mechanism make it a fantastic choice for video filming.

The FUJIFILM X-T3 has a 3" LCD screen that can rotate up to 180 degrees, so you can view your footage from any perspective. The camera's small HDMI port also allows for direct TV watching, live streaming, or both.

This book will teach you how to use the FUJIFILM X-T3 to record videos and how to view them afterwards. Its user-friendly design makes it easy for anybody to pick up and start making professional-grade films right away. We'll also talk about how to improve your recordings so that others may use them. Having read this chapter, you should feel confident using your FUJIFILM X-T3 to create cinematic masterpieces.

To begin video recording with the FUJIFILM X-T3, you need just do three quick operations.

To start recording, turn on the camera by pressing the power button.

Step two: use the rear control dial to enter Cinema Mode.

Finally, start filming by turning on the camera's video record button.

When you've completed recording, you'll be sent to a screen where you may view and/or save your video.

Movie viewing

The FUJIFILM X-T3 mirrorless digital camera can take high-quality photos and record high-quality audio and video. Its versatility makes it a great device for viewing movies. The 3" LCD display allows you to examine every detail of your recordings. Easy sharing of media with friends and family is made possible with in-built Wi-Fi and Bluetooth.

The FUJIFILM X-ability T3's to capture 4K footage gives the impression that you are right there in the heart of the action when viewed on its massive screen. Superior image quality and other cutting-edge features distinguish the FUJIFILM X-T3, a powerful, state-of-the-art camera.

It's great for viewing and recording movies because to its high-quality imaging technology, which does both with pinpoint precision and vivid colors.

Whether you're just starting out in the world of filmmaking or photography, the FUJIFILM X-T3 is the ideal camera for you.

It records in slow motion, for creating stunning cinematic films, in burst mode, for capturing a large number of photos in a short period of time, and in 4K, for producing amazingly crisp and realistic pictures.

The X-many T3's helpful features include a touchscreen interface, a tiltable lens mount, and support for a wide range of lenses and accessories.

Photo taking

The FUJIFILM X-T3 is regarded as one of the greatest cameras currently available. Its small form factor and low weight make it a convenient companion on vacation, and the camera's 24.2MP APS-C sensor captures stunning images with remarkable detail and color fidelity.

It has a quick and accurate focusing mechanism, an LCD screen that tilts for easy framing, the capability to record 4K video, and an electronic viewfinder that is integrated right in.

If you're a photographer who doesn't want to break the bank but yet wants to take pictures of good quality, the FUJIFILM X-T3 is a great option.

The camera boasts an APS-C sensor with 24.2 megapixels of resolution, an ISO range of 100 to 25600, and a wide selection of creative still and video picture shooting modes. It might take up to four hours to fully recharge the NP-W126S battery pack needed to power the FUJIFILM X-T3.

Shooting Mode Selection

As a result of its user-friendly design, the FUJIFILM X-T3 is a great camera for amateurs and professionals alike. The camera's many shooting modes allow for optimal results in every setting. Program Auto, Aperture Priority, Shutter Priority, and Manual are just some of the various accessible shooting settings.

Each mode comes with its own set of adjustments to help you get the best shot possible. The X-Shooting T3's Mode Selection button makes it easy to choose between the camera's several shooting modes to find the one that best suits your needs. This feature will be useful whether you're just starting off or have years of experience under your belt.

Automated

Shooting on "Auto" is frequent since the camera automatically adjusts the settings to the current lighting and subject matter. One of the numerous premium features it offers is Scene Recognition, which can recognize and optimize for different settings.

It also has a scene/object-specific adjustment mode called Advanced SR AUTO. With these aids, even a novice may shoot photographs that seem like they were taken by an expert. turn on the camera or enter a more involved shooting mode.

You may adjust the camera's settings for every situation or subject with the sophisticated SR AUTO mode. By using these presets, you may capture beautiful images without having to fiddle with the more advanced settings.

AE programmed.

The Fujifilm X-Program S10's AE (Auto Exposure) option was developed to streamline the photography process. Without having to manually fiddle with each parameter, it lets you choose the optimal exposure for the scene you're capturing.

Photographers who don't want to spend a lot of time fiddling with their equipment might benefit greatly from using this mode.

In Program AE mode, you may adjust shutter speed, aperture, ISO, and white balance, among other settings. Face identification, scene recognition, and automatic exposure adjustments are just a few of the more complex features it provides. All levels of photographers will find these elements helpful in producing professional-quality photographs.

Priority of shutter

In terms of power and versatility, the FUJIFILM X-T3 is hard to beat. Shutter Priority is a common photography option because it gives the photographer control over the shutter speed while letting the camera decide on aperture and ISO. You may capture fast-moving subjects or stop time in this mode.

By switching to Shutter Priority, you may take clear pictures with less blur and more precise control over the exposure. The

Shooting Menu is where you'll find the easy controls for the Shutter Priority Mode.

Recommended ISO and aperture settings for photography in Shutter Priority are as follows:

Speed of the shutter: 1/125 second to 1/500 second

ISO 100-200 Aperture f/13 or f/16

Priority of lens opening

The FUJIFILM X-T3 is a small mirrorless camera with a number of different shooting modes. The Aperture Priority photography option is a favorite because it enables the photographer to choose the desired aperture while the camera handles the shutter speed.

This setting is ideal for low-light portraits and landscapes with narrow depth of focus, as well as for limiting motion blur. It's also great for rapidly changing aperture settings without requiring any further manual adjustments. Taking stunning, one-of-a-kind photographs that stand out from the crowd is as simple as switching to Aperture Priority mode.

Manual Adjustment of exposure

If you're a photographer in need of an intuitive camera that yet gives you complete manual control, go no further than the FUJIFILM X-T3. The X-T3 has a Manual Exposure option where you may have complete control over the camera's exposure. The shutter speed, aperture, and ISO may all be customized in this mode to get the desired effect.

You'll have a lot more say over the final product of your photographs by using this mode, since it grants you access to other shooting modes including Program AE and Aperture Priority AE. The Fujifilm X-Manual S10's Exposure mode equips you with all the controls you need to take stunning pictures in any lighting scenario.

AF settings

The focusing system of the FUJIFILM X-T3 is quite sophisticated. The X-sophisticated S10's focusing mechanism allows you to take sharp pictures of your subject in a flash.

The camera's Face/Eye Detection AF and Tracking AF make it simple to take clear pictures of subjects that are moving quickly. The FUJIFILM X-T3 is a great option for photographers who wish to easily take clear shots because to its excellent autofocus technology and wide array of focusing modes, including Single Point AF, Zone AF, and Wide/Tracking AF.

Focus mode.

A camera with autofocus may be set to zero in on a specific subject, making it much simpler to get clear, focused shots every time. With this function, photographers of all skill levels may take sharp pictures without fiddling with the focus. With autofocus on, getting sharp pictures requires next to no effort. The FUJIFILM X-T3 is a great option for serious photographers due to its versatile focus settings.

Focus point selection.

Single-point, zone, and wide/tracking options for choosing the focal point are all included. This gives you the freedom to choose the lens that works best for any given scenario. The X-autofocus S10's mechanism makes it simple to take clear pictures every time.

In addition, it has the ability to recognize faces and eyes, so your subjects will always be in sharp focus. The FUJIFILM X-T3 is a great option for any photographer searching for a durable camera with exceptional focusing skills thanks to its sophisticated autofocus mechanism and many focus point selection settings.

Focusing using manual mode

The FUJIFILM X-T3 is equipped with cutting-edge autofocus technology, and it also has a manual focus option for photographers who want more creative control. As a result of its straightforward interface, you can quickly and accurately focus your photographs. The Fujifilm X-manual S10's focus settings will let you to take professional-quality landscape and portrait photographs every time.

The FUJIFILM X-internal T3's 3.7V lithium-ion battery permits up to 650 photographs in highest quality mode, or 200 shots with the electronic viewfinder activated. In addition to the USB port on a computer, an AC adapter may be used to charge the camera.

Tonal range

With its increased sensitivity, the FUJIFILM X-T3 is a digital camera worth considering. The camera's sensor is superior to those of competitors, allowing it to record brighter and more detailed images. This expands the image's tonal range, resulting in more nuanced colors and sharper differentiation between bright and dark areas. The camera's sophisticated focusing technology allows for rapid subject acquisition, even at high shutter rates. This enhanced sensitivity allows photographers to take breathtaking pictures in almost any lighting.

Meter settings

For this reason, the FUJIFILM X-T3 offers a wide variety of metering modes. For accurate exposure, the camera's 256-zone metering technology collects measurements at regular intervals.

There are three exposure metering modes including Multi, Spot, and Average. There's also an Advanced SR Auto setting that determines the best metering mode for each photo. The metering system of the FUJIFILM X-T3 is so exact that it makes it a great choice for any photographer who wants to shoot professional-quality photos.

By using these controls, you may rapidly change the exposure of a scene to suit the available light and the impression you want to convey.

You can get an accurate reading of the scene's overall illumination using the Multi Metering mode since it employs a vast area of the sensor to do so, but in certain cases, it may cause your images to be underexposed or overexposed, depending on the scenario and the subject matter.

Spot metering allows the camera to concentrate on a tiny area of the frame and determine the proper exposure based on the subject matter that is directly in front of the flash.

Last but not least, while using Average Metering, the camera uses an average of the viewfinder's brightness readings to determine how much light will be captured in each shot.

Locking exposure

Locking in your shot's focus and exposure using the X-Focus/Exposure S10's Lock function ensures that your photos will always turn out beautifully. When photographing in low light or attempting to catch fast-moving objects, this function becomes invaluable. Insist on perfect results every time with the Focus/Exposure Lock.

Fujifilm's X-T3 is a stripped-down camera that packs a lot of bang for its buck. With its manual controls and selection of three separate filters, the X-T3 gives photographers a wide range of leeway to express their individuality through their work. A Wi-Fi function lets you use your phone as a remote control when shooting images or films, and the camera's dual LCD displays give you even more creative control over your shots.

HDR settings

The FUJIFILM X-T3 digital camera uses cutting-edge HDR (High Dynamic Range) photographic technology. Thanks to this innovation, the camera can take pictures that are both beautiful and technically advanced. It can shoot images with better dynamic range by simultaneously recording numerous exposures and merging them into a single picture. The X-T3 can shoot in very low light with to its outstanding ISO range of up to 51200. Its high-tech focusing technology allows it to zero in on any subject with lightning speed and precision.

The X-user-friendly T3's UI and straightforward settings make it simple to take professional-quality photos in any lighting condition.

Settings for Panorama

The LCD screen on this camera can swivel, and it also has an electronic viewfinder, so framing your photographs is a breeze. You may take stunning panoramic shots with a single click of the shutter when using the Panorama mode.

If you want to take breathtaking panoramic photos of nature or the city, the FUJIFILM X-T3 is an excellent option.

The FUJIFILM X-T3 is a versatile camera that performs well in a wide range of environments, from a rock concert to your living room.

Many of the camera's settings are preset for use in a variety of situations. You may further personalize your images with a single click of the shutter and a selection of manual settings.

Different exposures

The Multiple Exposure mode makes it simple to produce stunning visuals by splicing together many photographs. There has never been an easier way to practice your photography abilities and create breathtaking, one-of-a-kind shots than with this feature.

The FUJIFILM X-T3 is loaded with tools to help you shoot the finest shot imaginable, including a tilting LCD screen, 4K video recording, and more. If you're looking to capture 4K video, the Fujifilm X-ability S10 is a great choice.

This camera can take photos and capture videos with outstanding detail, clarity, and color contrast. With the photo/video mode, you may do things like take panoramic photographs, use slow-motion effects, and record time-lapse videos.

A 4K mode option on the FUJIFILM X-T3 that works for the shots you want to take is likely. This cutting-edge camera is a great tool for those who like working with filmmakers and photographers. For a more tailored experience, the FUJIFILM X-T3 offers a slew of extra capabilities and customization options.

Using flash

A flash is an essential photography accessory. Making beautiful images in low light or adding drama and mystery to any environment is possible with these tips.

Because to its many helpful features, the FUJIFILM X-T3 is an excellent camera for capturing stunning shots even in dim lighting.

The X-high S10 always takes great photographs because to its high ISO sensitivity, fast shutter speed, and powerful flash system. Using a flash in a photograph has never been easier.

The FUJIFILM X-T3 is the perfect camera for every photographer, whether they are just starting out or are seasoned pros. The camera's APS-C sized, 16 megapixel sensor, in conjunction with the X-Processor Pro Engine and EXR Processor II, enables quick focusing, precise color reproduction, and little noise at high ISO settings.

Some of the many great features of this camera include the 3" Excellent viewing in all lighting on the TFT LCD panel, the various viewfinder options (including an electronic level), the built-in HDR setting for beautiful images, and the ability to record in full HD.

Image taking and replay.

The FUJIFILM X-T3 is a fantastic camera for both novice and seasoned photographers because to its large APS-C sensor, fast focusing, and 4K video capture. Because of the camera's intuitive interface, taking and reviewing pictures is a breeze.

The many benefits include a large APS-C sensor for high-quality images and a host of additional features.

-High-performance, cutting-edge technology and software for instant autofocus.

Because to its user-friendly interface, it may be used by photographers of all ability levels.

- Fantastic features for professional photographers and casual viewers alike, including timelapse mode and 4K video recording

Photo taking

The FUJIFILM X-T3 is an excellent option for photographers because it features a big

APS-C sized sensor, which provides both good image quality and broad dynamic range, and because it is compact, lightweight, and simple to operate.

The electronic viewfinder and 4K video recording capabilities make the FUJIFILM X-T3 an excellent choice for any photographer who wants to easily capture stunning images.

Taking photos with the FUJIFILM X-T3 is straightforward and requires only a few motions thanks to the camera's intuitive interface and powerful sensor and image processor.

Learn how to use the FUJIFILM X-T3 to shoot professional-quality photographs with this helpful tutorial.

You may modify your ISO or the amount of light utilized for your shots if you want to capture pictures in various situations, but first you need to switch on your camera and make sure it's completely charged.

Once you've mastered the fundamentals, you can start taking pictures by pressing the camera's power button.

To link your camera to a mobile device, tablet, or computer, activate its Wi-Fi function.

Select "On" in the Wi-Fi configuration menu to enable wireless device connection to the FUJIFILM X-T3.

Second, position the camera so that the lens is pointing in the direction you wish to shoot.

Third, launch the Fujifilm Camera Remote app on your smartphone and tap "Set" to establish a connection between it and your camera.

Photo viewing

The FUJIFILM X-T3 is a powerful and versatile camera that lets you take stunning photos in any situation, thanks to its user-friendly interface and wide range of features.

This chapter will provide you with some tips and steps on how to view your photos taken with the FUJIFILM X-T3, including viewing in playback mode, using the LCD screen, and adjusting the brightness.

Playback mode viewing

Using the camera's replay mode, you may examine the pictures you've already shot.

To start recording or viewing your footage:
1. Press the power button to switch on your camera, or press the play button to enter playback mode.

You may scroll through your images in a thumbnail index by pressing up or down (the index will only appear if you have taken more than 24 photos).

Using the H/press and left or right, you may navigate through your photographs from top to bottom of the screen.

If you have more than 24 photographs, you may access the index display by pressing the control key, and then using the left and right arrow keys to go through your photos.

When you press the joystick up or down, you may scroll through a series of images in an index format (the last photo will be shown in a zoomed-in thumbnail).

To leave this mode, hit the power or play button again.

Photo deletion

Delete images from your FUJIFILM X-T3 with some effort, but be prepared for some complexity. Here, we'll go through the process of erasing photographs from your FUJIFILM X-T3 so that only the finest shots are preserved.

The intuitive design of the FUJIFILM X-T3 makes it simple to delete photos.

With the help of this article, deleting photos from your FUJIFILM X-T3 should be a simple, allowing you to make better use of the camera's storage space.

To begin, press and hold the shutter button to turn off the camera; this will also safeguard your photos by keeping them in the camera's buffer even after the power is turned off. Just pressing and holding the camera's power button once more will turn it back on.

Second, pick Camera mode from the Menu/Setup and then press the Mode button. To access the last mode, choose "2," and then "3," a little black box.

Hold down the camera's BACK button until the "Red Cross" screen displays, as seen in the image below (Step 3).

Your new photo should appear in the upper left corner of the screen after around 2 seconds, followed by a "1."; at this point, you may either take another photo or look at the one you just shot, depending on your requirements. When the night vision functions of your camera are activated, a brilliant orange circle will appear at the camera's focal point, with a kind of haze surrounding it.

But, you can still take a picture and look at it in the Pictures app to see what the camera catches at night if you don't utilize this mode.

Shooting Modes Selection

The FUJIFILM X-T3 excels in terms of its shooting possibilities. A lot of helpful features that make picking a target easier and faster are included. The camera's user interface is straightforward, and it comes with a wide variety of lenses and excellent focusing. Also, it has a powerful image processor capable of handling images of any resolution. The FUJIFILM X-T3 allows you to shoot high-quality images for any purpose.

Setting for image quality

The FUJIFILM X-T3 offers a wide range of adjustments, from the simple to the sophisticated, to ensure you get the best shot possible. The intuitive design and features of the application make it simple to adjust the settings for optimal image quality. From the white balance and color saturation to the corner sharpness, you'll be able to adjust anything with a single camera. Whether you're a landscape or portrait photographer, the FUJIFILM X-T3 has all the image quality settings you could possibly need.

Image size

Among its many other high-quality photo features, the FUJIFILM X-T3 lets you choose your image resolution. This feature allows you to quickly and easily alter the size of your photographs while maintaining their original quality. In this book, you'll learn how to use the image size settings on your FUJIFILM X-T3 to take pictures at the best possible resolution.

Quality of image

In order to capture stunning photographs, the FUJIFILM X-T3 is an excellent choice of camera. To maximize the quality of your photographs, this camera offers a variety of picture quality options. Choose from a wide range of options to get the highest possible picture quality in either RAW or JPEG shooting modes.

Using this function, you can make sure your final photographs are clear and colorful, without any blurring or other artifacts. The X-T3 also offers a film simulation mode that mimics the appearance and feel of traditional film stock while taking images.

RAW filming

With the inbuilt Raw recording mode, you may snap photographs of the highest possible quality. This will help you get the clearest and most detailed shots possible.

You may use this feature to fine-tune your images by modifying aspects like ISO, white balance, and color depth. It also records in a variety of formats that may be easily processed in post-production, including as JPEG and TIFF.

High-powered and packed with unique capabilities, the FUJIFILM X-T3 mirrorless camera lets you simulate the look of different film stocks by tweaking the camera's color profile, contrast, and sharpness. Photographers may now take shots that are visually and conceptually extremely close to their film-based predecessors. Photographers using the FUJIFILM X-T3 may experiment with different styles and create stunning images with a retro feel by adjusting the camera's image quality settings.

The ability to mimic different films by altering the camera's color profile, contrast, and sharpness is one of the most intriguing features of this high-powered mirrorless camera. This allows photographers to mimic the look and feel of traditional film photography. Photographers using a FUJIFILM X-T3 may experiment with new techniques and create stunning photos with a retro feel by adjusting the camera's image quality settings.

Simulation of film

Shooting with the FUJIFILM X-T3 digital camera is like using film in many ways, and the two mediums were designed to feel quite similar to one another. Unique among its type, it allows photographers to simulate the appearance of classic Fujifilm films like Velvia, Astia, and Provia. The X-T3 has a number of post-processing tweaks and color presets to help you get the look you want. Using these components in tandem, photographers may create effects that are reminiscent of classic film.

Black and white

Capturing photographs in monochrome is now possible, and all of the camera's other settings and functions will still apply. Use filters and effects to give your monochrome photos a one-of-a-kind feel. The FUJIFILM X-T3 can help you achieve both traditional black and white photos and more experimental color combinations.

Effects with grain

If you're a photographer who loves the grain look, the FUJIFILM X-T3 is the camera for you. With this function, you can give your photographs a one-of-a-kind appearance by applying a special effect. The grain effect of the FUJIFILM X-T3 may be adjusted from very faint to very dramatic. Focus peaking and facial identification are just two of the helpful extras that come with this camera. The FUJIFILM X-T3 is a wonderful pick for any photographer searching for a simple method to add interest and character to their images thanks to the camera's amazing grain effect capabilities.

Color aberration

This filter was created to make your photographs seem more alive and vibrant by bringing out the full range of colors contained within them. Even more so when photographing landscapes or cityscapes, this effect may help you produce really unforgettable photos. The Color Chrome Effect allows you to give your photographs a unique look and feel. In addition to its 4K video recording, electronic viewfinder, and image stabilization technology, the FUJIFILM X-T3 boasts a number of additional useful features.

Using white balance

The ability to control the white balance of your camera is crucial for getting true color reproduction in your photographs. It's the method used to make whites in photos seem white rather than off-white.

Both the manual and automatic white balance settings on a camera might assist with this. Photographers may prevent unwanted color casts and capture true-to-life hues by adjusting the white balance. or unwelcome hues

The White Balance setting on a camera makes all the colors seem the way they would in perfect lighting. Depending on the lighting conditions at the time and place of the shot, the results will vary if this parameter is altered.

Color range

The dynamic range of a sound system is measured by the difference between its loudest and quietest playback levels. Because it dictates how effectively a system copes with varying sound levels, dynamic range is a crucial consideration in audio production and engineering.

The bigger the dynamic range, the wider the range of audible volumes that may be faithfully represented. Systems with a higher dynamic range may faithfully reproduce softer tones, while those with a lower range may distort quieter sounds.

Curves of tone

The tone curve is used to adjust the exposure levels in an image. A histogram is a graphical representation of an image's tonal range. Adjusting the tone curve allows photographers to change the image's contrast, brightness, color saturation, and hue. Manual or automated adjustments to the tone curve are possible in image editing programs like Adobe Lightroom and Photoshop. This tool is great for helping photographers increase the dynamic range and color saturation of their photographs. At its simplest, a tone curve is a straight line, since a single hue has just one tone.

Usually, shadows will initially cast a downward slant over the center of the graph. Then, the shadows and highlights are represented by two huge peaks, one on each side of the dips. With these levels of contrast, the image is at its most visually arresting and interesting. The photographer may then decide whether to leave the contrast as is or boost it at either peak.

Dynamics

The FUJIFILM X-T3 may be purchased in a number of different colors, and it is a high-quality digital camera. The camera's color settings were made with versatility in mind, so that you may take beautiful landscape or portrait photos. The X-T3 may be set to one of three distinct color modes: Natural, Vivid, or Monochrome. Each setting has its own personality and may be customized to achieve a certain aesthetic in any given photograph. Photographers may use the X-flexible S10's color settings to capture great shots with either vivid hues or muted tones.

Color is a potent psychological and social manipulator. As such, it is an essential component of visual communication, serving to transmit meaning, establish atmosphere, and elicit response. Products and services may be set out from the competition with the use of color. Businesses may improve their branding and marketing with a deeper grasp of the psychological effects of color.

Clearness of image

The FUJIFILM X-T3, for instance, is a groundbreaking camera because it brings together the advantages of digital and analog photography in one convenient package. With its 26.1-megapixel X-Trans CMOS 4 sensor, improved picture processor, and 3.0-inch, 1.04 million-dot tilting LCD panel, users can capture breathtaking stills and moving footage. In addition, the camera's straightforward interface makes it suitable for both amateur and professional photographers alike. The Fujifilm X-lightweight S10's and small form factor make it an ideal camera for taking on the road. It is one of the most powerful cameras available today, with features including 4K video capture, autofocus that follows the subject's eyes and face, and a burst rate of up to 30 frames per second.

Noise reduction

The High ISO NR (Noise Reduction) function makes the FUJIFILM X-T3 a very useful camera, allowing for sharp and clear photos even in dim lighting. With this function, the camera is able to minimize noise and capture crisp, almost noise-free shots, even at very high ISOs. The X-T3 is a superb option for photographers of all skill levels because to its quick focusing, user-friendly touchscreen, and extensive collection of creative controls. The FUJIFILM X-T3 is equipped with High ISO NR technology, allowing for stunning photographs to be taken in low-light settings.

Color gamut

The FUJIFILM X-T3 is an electronic viewfinder small camera. An APS-C sensor with 24.2 megapixels, covering both the Adobe RGB and sRGB color gamut, is included.

The FUJIFILM X-T3 is a great option if you're seeking for a camera that consistently produces high-quality photos. It has a high-

resolution APS-C sensor with 24.2
megapixels that can capture images in
Adobe RGB and sRGB color spaces. The
electronic viewfinder of the FUJIFILM X-
T3 allows the photographer to keep their eye
on the subject at all times and achieve sharp
focus with ease.

Settings for autofocus.

Improve the quality of your photos in any
environment with the FUJIFILM X-T3 and
its AF Setting. The AF Setting allows the
user to focus the camera's lens on a specific
part of the image. Even in low light,
photographers can swiftly snap clear images
thanks to the camera's fast focusing. You
may activate or deactivate autofocus through
the camera's touchscreen or a dedicated
button. It's a great tool for recording video
and taking still photographs of people,
landscapes, and all kinds of animals.

Main Point

A 24.2-megapixel APS-C CMOS sensor and
a 3-inch LCD screen are included in the
FUJIFILM X-T3 digital camera. It includes
an ISO range of 100-25600, a continuous

shooting speed of 16 frames per second, and the ability to record 4K video at 30 frames per second. The X-T3 from Fujifilm is an affordable first-level camera with a wealth of practical enhancements. This is an excellent choice for people interested in photography but unable to make a significant financial commitment right now. Cameras with a wide field of view and a modest price tag that can focus on a wide variety of subjects.

Toggle Focus Mode

Using the FUJIFILM X-new T3's Focus Mode, users may lock focus on a moving subject and have the camera track it.

Focus Mode is ideal for capturing action or animals. Focus Mode saves you the time it takes to manually find your focus point by allowing you to set it and then let the camera find it.

The "AF mode" on the Fujifilm X-autofocus S10 is designed for those who only want a quick and easy way to take great photos. You may take pictures with this camera without having to fiddle with the settings by selecting the "Auto" option. Moreover, there are:

A Facial Recognizability Capability

Prioritization based on facial recognition. Totally apt for a photo shoot

The 4K Video Standard Is the Future

Take full use of the X-high T3's definition video recording capabilities. Capturing video in tones ranging from natural to stunning is made possible by a variety of frame rates, resolutions, and color profiles.

Autofocus mode.

The AF point display is new to the FUJIFILM X-T3, making it a high-end small camera. The Display and viewfinder are both touchscreens enabled.

The AF point display on the Fujifilm X-unique S10 aids with pinpoint focusing. Users may zero in on their intended objectives with the help of a distance indicator shown in meters.

The FUJIFILM X-T3 will retail for $699 upon its September 2019 release.

Showcase of AF points

Users of the FUJIFILM X-T3 digital camera have access to a plethora of settings and configurations. In terms of sensor size, it's identical to the Fujifilm X-APS-C T2's unit (24.2 megapixels). The FUJIFILM X-T3 is an excellent choice for anyone in the market for a camera that can reliably provide professional-grade results.

Settings for shoot

The digital Fujifilm X-T3 camera is packed with cutting-edge features. At 24.2 megapixels, the APS-C sensor is on par with that of the Fujifilm X-T2.

Go no farther than the Fujifilm X-T3 if you're in need of a camera that reliably delivers professional-grade results.

Where the action is taking place

The Fujifilm X-T3 is the company's entry-level mirrorless camera, and it comes with a fixed 24-millimeter lens. Its small sensor and broad aperture range (f/2.0 to f/16) make it an excellent choice for photography in dim conditions.

If you're just starting out as a photographer but want to invest in a high-quality camera without breaking the bank, the Fujifilm X-T3 is an excellent choice.

The 18-55mm lens may be swapped out for a longer focal length option of between 28 and 85mm. A telephoto zoom lens with a focal length range of 70-300 millimeters is also available for use with it.

The camera's auto exposure and four-megapixel focusing sensor may be adjusted either manually or automatically. Fujifilm's X-S10 may be used with fully manual control over shutter speed, aperture, and white balance.

Changing the Filter Settings

The Fujifilm X-T3 is a compact and lightweight camera, making it ideal for travel. Just two of its many remarkable capabilities are the ability to capture 4K video and RAW data.

This camera is ideal for photographers who are always on the go. Yet, this camera is unique in that it lacks a back control dial and buttons. It's annoying to have to take your eye from the viewfinder while you fiddle with the shutter speed, aperture, or ISO.

Fujifilm's filter setting software may be used to solve this issue and provide you more creative control over your photography. If, for instance, you're shooting at night with an ISO of 100, you may use an ND8 filter to reduce light while still achieving passable picture quality.

Locator Sport

The electronic viewfinder makes the compact Fujifilm X-T3 suitable for anything from casual photography to capturing dramatic moments.

While the f/2.8 lens and 24.2MP APS-C sensor produce high-quality shots in bright light, they struggle in low-light situations due to the lens's smaller maximum aperture when compared to, say, the Sony A9 or the Canon EOS R, which have maximum apertures of f/1.6 and f/1.5, or the Nikon D850, which has an aperture of f/0.95.

Preset timer for one's own use.

Fujifilm's latest digital camera, the X-T3, has an integrated self-timer. An APS-C sensor with 16.3 megapixels, 4K video recording, and an electronic viewfinder are just some of its features. The X-T3 is a brand new model for Fujifilm this year.

The built-in self-timer makes it easy to get great selfies and group photos. Four-K video recording and an electronic viewfinder are two other standout features.

The X-T3 from Fujifilm is a high-resolution digital camera with a 16.3-megapixel APS-C sensor that is capable of recording 4K video. It includes an electronic viewfinder that can be replaced out with a different Fujifilm viewfinder kit and can record 4K video at up to 60 frames per second.

The Fujifilm X-features T3's are extensive, including burst shooting at up to 9 frames per second, phase detection auto focus, a self-timer, RAW capture, intervalometer, movie creator, time lapse mode, color effects for stills and films, and more.

Automatic firing at preset intervals

The Fujifilm X-T3 has an interval timer feature. This feature allows you to set the camera to automatically take pictures at regular intervals, relieving you of the need to continually press the shutter button.

The Fujifilm X-T3 has a timed interval shooting option. This is the best choice if you want to take pictures without having to continually press the shutter button. In addition to taking a sequence of stills from a moving subject, the interval timer shooting mode may also be used to combine many images of a moving subject into a single image.

When time is of the essence, but you still want high-quality photos, the Fujifilm X-T3 is the camera to have. Beautiful low-light shots are possible thanks to the camera's 24-megapixel APS-C sensor, and the availability of manual controls provides photographers more leeway to express their own creativity.

Measure the brightness of an item with the use of a technique called photometry. A light meter is used to take readings, and the information gained from those readings is then utilized to build an image file.

The Fujifilm X-T3 is well-liked among photographers because to its built-in photometry function. A light meter or a tripod aren't necessary for quick images like this.

The following are just some of the many fields in which photometry finds use:

Taking photos in low light, when editing software like Lightroom and Photoshop can be used to boost contrast and depth of field; photographing subjects that are notoriously tricky for cameras to focus on, like flowers; photographing people who are moving quickly while still wanting to capture motion; photographing objects in motion.

Mode of a shutter

The APS-C sized Fujifilm X-T3 has 24 megapixels and an electronic viewfinder. While not in use, the lens of the camera may be retracted into the body.

Diminished flickering

The glow of our electronic devices might be an annoyance. Since screens are so ubiquitous in the modern digital age, this is crucial. If you use a flicker-reduction filter, you may reduce the strain on your eyes caused by staring at a screen for long periods of time.

The number of times a bulb flickers may be reduced by as much as 80% with the help of the flicker reduction filter on the Fujifilm X-T3 camera. It uses both an optical and an electrical low-pass filter to reduce the perception of flickering light without compromising color fidelity or gamut.

Settings for flash

The flash mode of the Fujifilm X-T3 helps produce photos with a more natural look and feel. This is due to the fact that it may be used in manual mode for more creative control and has adjustable settings.

This camera has an electronic viewfinder, so you can check in on your subject without taking your eyes off the action.

Tweaking the flash function

The X-mechanical T3's shutter speed of 1/4000 sec. allows for sharper photos than those of competing cameras. This feature allows users to take good pictures despite low light or motion. This camera has an impressively high ISO range, usable from 100 all the way up to 16,000. It makes it usable even in low-light conditions.

Subtraction of Red Eye

Red eye is an uncomfortable and often humiliating condition that affects many people, especially in low light. The Fujifilm X-T3 has a built-in red-eye correction feature, which may be used to remove red eyes from photographed subjects.

Preventing a Bloodshot Appearance: We've all felt the pain of red eyes, especially in low light. The Fujifilm X-built-in T3's automatic red eye correction makes quick work of fixing images damaged by squinting subjects.

Lamp Configuration for LEDs

The Fujifilm X-T3 digital camera has a built-in LED light option. A photograph's white balance and color temperature may be quickly adjusted by the photographer.

The Fujifilm X-T3 allows for a wide range of changes to be made to the shooting settings in the present time. The camera can capture images from as little as 1/4000 of a second to as much as 60 seconds, and the ISO can be adjusted from 100 all the way up to 12,800. The electronic viewfinder has a 0.5-inch screen, and the sensor is a full 16. The Fujifilm X-T3 is available in either black or silver.

The new Fujifilm X-T3 is an impressive camera with a lot to offer and a distinctive design. This camera is great for shooting movies since it has a built-in electronic viewfinder, a built-in microphone, and a grip that can be adjusted to accommodate hands of varied sizes.

Settings for film

The Fujifilm X-movie S10 has a mode that allows you to take both still images and video. This versatile location is ideal for photographing everything from landscapes to sporting events.

Compact and capable of recording 4K video and beautiful still photographs, the Fujifilm X-T3 is a must-have for any photographer's kit. Both the electronic viewfinder (magnification 0.39x) and LCD screen (1.04x) make it easy to get a good look at your subject before taking a picture.

This camera's capabilities are extensive, including but not limited to 4K video recording at 30 frames per second and 1080p video recording at up to 60 frames per second, continuous shooting speed of up to 10 frames per second, ISO 100 - 12800 range for stills, ISO 6400 range for video recordings, support for the RAW file format, high definition movies (HDMI out), and weather.

Information Storage Format

A video file holds all the information necessary to play or edit a video. Details such as codec, bitrate, resolution, frame rate, and so on are included in the technical specs.

Video file formats have several uses, such as internet video streaming and camera-based video capturing. At now, MP4, AVI, and MKV are the most widely used video file types.

Placement of sound system

The Fujifilm X-T3 digital camera has cutting-edge noise cancellation and sound processing. In order to take the finest image possible, the camera will adjust its settings automatically based on what it detects.

Quality of image

The picture quality may now be customized on the Fujifilm X-T3 camera. This option, accessible in several shooting modes, allows for more control over final picture quality, particularly when working with the RAW file format. The picture information is updated to reflect the image quality you selected here, and it will be available in other applications that can read RAW files. This camera has a JPEG fine option by default, but you may change it to any of the others listed below.

In comparison to JPEG medium or standard, the file size of JPEG basic/Better (Compressed) is less, and you have more control over the colors in your image. You may acquire high-quality images with a file size that's appropriate for online publication by selecting the JPEG medium quality setting rather than the lower-quality Fine mode, which delivers greater resolution, more clarity, less noise, and smoother gradations.

The JPEG standard is the best choice for online publishing and emailing because of its compact file size.

Compared to JPEG medium and standard, JPEG fine has a reduced file size and more precise color management. The Fine setting improves upon the Standard setting in every conceivable way.

Film Simulation

Fujifilm's newest camera, the X-T3, is part of a long series of high-quality devices. It has a user-friendly interface and a mode that mimics the behavior of various films. This setting makes it simple to snap images with a classic vibe or to switch to high-contrast black and white.

Dynamic Range

The dynamic range of a camera is its ability to record a graduation from very bright to very dark areas of a scene.

Photographers value dynamic range for its ability to capture a wider variety of tonalities, from deep blacks to bright whites. It's not only for landscape and portrait photographers, but also for those who dabble with HDR.

The Fujifilm X-T3 has a dynamic range of over 11,000:1 and is one of the most reasonably priced cameras in its class. It's the key to unlocking your full photographic potential!

Tone Curve

Fujifilm improved the X-T3 camera by including a new Tone Curve function. It's a great way to give your shot a new vibe, whether you want to make it more dramatic or charming. Adjusting the curvature of a photograph in this way lets you fine-tune the aesthetic of your final product.

Sharpness

A camera with a low noise level like the Fujifilm X-T3 is a must-have for any still or moving image maker serious about their craft.

The Fujifilm X-T3 has remarkable crispness, making it ideal for both stills and video.

High ISO NR

Historically, photographers were required to utilize a moderate ISO level to get acceptable results. The Fujifilm X-T3 makes it simple for photographers to snap pictures in low light or with a high ISO.

The Fujifilm X-T3 is a reasonably priced camera that, because to its High ISO NR mode, can capture clear images even in dim lighting.

The camera's wide-angle lens and anti-shake mechanism make it simple to capture expansive environments with little distortion.

If you need to photograph a large group of people but are not quite ready to commit to a DSLR, the Fujifilm X-T3 is the camera for you.

Settings for audio

In terms of sound, the Fujifilm X-T3 offers three different choices. There are three distinct categories, and these are the Normative, Imaginative, and Cinematic. In most situations, Standard is the best photography mode to employ. While taking pictures of people or landscapes, use the Creative mode. All shooting should be done in Movie mode.

The default volume levels are "Normal," "Music," and "Voice." Vocal communication was developed specifically for use in busy, loud settings where it would be difficult for others to hear them.

This is the equipment you have if you desire harmony in your landscape or portrait shots.

In this mode, video recording uses the system's default audio settings, which don't provide you many options.

In the innovative mode, you are able to adjust the loudness of any background noises and choose the delay before the music starts playing between photos. While in this mode, you may adjust the volume to either the standard level or the music volume.

The audio settings in the movie mode are more flexible. The volume of the background noise and the length of time that passes between shots before the audio starts playing again are also controls that the user has full control over.

Arrangement for a Microphone Jack

When using an external microphone, the Fujifilm X-T3 is equipped with a jack for such a connection. The camera's controls are straightforward and easy to learn, but there are a few significant areas in which they differ from the controls on other cameras.

Changing the mic settings is done under the "Recording" submenu of the camera. You may adjust the volume of your recorded voice here.

You may modify the volume of the microphone input by dragging the "Level Switching" slider, which can be found in the "Recording" submenu. You can increase the amount of sound picked up by your microphone by turning this up, and vice versa.

The Microphone Ring Gain setting allows you to adjust the amount of gain given to any microphone plugged into the camera's mic port. The ring gain button may be used only when a microphone is being used to record; the phrase "ring gain" is explained.

Wind Shield

The Fujifilm X-Wind S10 Filter might help reduce wind noise and enhance your outdoor photography.

The Fujifilm X-T3 is an excellent choice for amateur photographers just getting started. It may be used as a regular camera or as a selfie camera thanks to its 3" LCD screen, 100% coverage electronic viewfinder, and 16 megapixel digital viewfinder.

In every setting, the Fujifilm X-T3 will produce photos and video that are worthy of a professional portfolio. It records high definition video at up to 12 frames per second, and 4K video at up to 8 fps.

The Low-Pass Filter

The low-cut filter is one kind of filter that works by reducing the volume of the audio's highest frequencies. This is the method to use if you need to capture a conversation or a speech while maintaining high quality sound for later playback.

Cable: Long, thin copper wires or rubber cords are used in the electrical and communications industries to transmit signals from one site to another.

Featured prominently: The best place for a center channel speaker is in a large area like a theater, auditorium, conference room, or classroom.

The music will be played in front of you from the center speaker, but it will seem to be coming from above your head if you have surround speakers installed. While sound may come from any direction whether you're at home or outside, it can only originate from below in a theater or a conference room (center).

Different codes

The Time Code Setting option on the Fujifilm X-T3 is quite helpful. This will help them take better pictures and keep their creative juices going.

When shooting progresses, the time code will be shown prominently.

There are two ways that users may enter the time code while taking a photograph.

Two methods are available for assigning timestamps to photographs: either by entering a unique timecode for each shot, or by entering a single timecode that will be split into many codes once a sequence of shots has been taken in fast succession.

This feature is used by photographers and filmmakers in the industry to keep an eye on their work and make sure they don't miss any important shots or situations.

Replay options.

Fujifilm X-T3 owners may take up to a hundred successive images in a row and choose the best one afterwards with the help of the camera's Replay Menu.

Furthermore, it may be utilized throughout shooting to prevent missing any important shots.

Unlike conventional photography, the new Replay Menu technique does not need any further actions after the shot has been taken, such as editing or filtering.

controls similar to those of a digital single-lens reflex camera, including ISO, f-stop, and shutter speed.

Viewfinder optics with 360-degree field of view and autofocus following capability.

You can operate the camera from afar thanks to its built-in Wi-Fi, Bluetooth, and GPS.

Extending the AF Tracking Range
Accurately and Capturing Faces (up to 10m)
All of the following modes (Superior Zone
AF, Wide Area AF, Centered Object AF,
Spot AF, and Single Point Object AF) are
available.

Display of replay

To check out your shots before you snap
them, you may use a feature called the
Playback Display.

It displays the image in the same way the
viewer would see it, giving you an
opportunity to make adjustments before you
actually take the picture.

Selfies in peace, examining your trip
images, and keeping up with your friends'
latest status updates are just some of the
many uses for the Playback Display.

The purpose of the Playback Display is to
facilitate the taking of high-quality
photographs for the purpose of sharing them
with loved ones.

Viewing Pictures

A Look at the Recording's Playback Screen For 2018, Fujifilm has released the X-T3. The 3.2-inch LCD screen may be utilized for viewing videos or browsing through images.

It's Playing Indicator A completely new Fujifilm product, the X-T3, was released this year. It has a 3.2-inch LCD display, making it ideal for viewing movies or browsing through photos.

The Fujifilm X-T3 is a digital camera with a viewfinder for examining shots before they are permanently stored. This is a one-of-a-kind camera since it has a built-in display.

It's convenient since it eliminates the need for the user to carry along a separate device in order to check their images.

The Fujifilm X-screen may be used for a variety of purposes. S10's Rather of holding up the camera and potentially disturbing the snap, the screen may be used as a viewfinder for situations when doing so would be distracting, such as while photographing children.

Replay options.

The most recent picture or video recorded may be accessed directly from the camera by using the Playback Menu, a new addition to the Fujifilm X-T3.

When deciding whether or not to preserve your most recent photo, you may watch it again quickly and easily from the Playback Menu.

Users may also instantly remove undesired photos and movies from the camera using the menu.

To enter this menu, hit the playback button, which is located on the top left of your camera.

Real-time preview of your live video is possible through the Live View Display. When shooting in rapid succession, this mode is a great aid.

To reach this menu, hit the playback button, located on the top left of your camera.

Options for Slideshow

The menu on the Fujifilm X-Playback S10 makes it easy to create stunning photo slideshows.

The presentation might be based on your own photo collection, which you can import.

You can make your slideshow really one-of-a-kind by adding narration and musical accompaniment.

The Fujifilm X-T3 is a mirrorless interchangeable-lens camera that is both compact and sleek.

There is a strong zoom lens in addition to the wide-angle main lens, so you may use it for landscapes and close-ups.

The camera can blend two photos taken at different exposures into a single HDR shot, and it offers 26 post-processing effects meant to add creative flare.

The S10's built-in Wi-Fi with X allows you to share your best photos on social media sites like Facebook and Instagram.

The camera's built-in GPS system enables you to easily tag any shot you take with its precise position.

If you're looking for a great camera at a reasonable price, the X-T3 is a great option. Go no farther if the price tag is a primary factor in your camera purchase.

The f/2 aperture, 2x optical zoom, and 16 megapixels are all impressive features of the X-T3. It also has built-in Wi-Fi and GPS, so you can quickly add location information to your photos.

The camera is HDR-compatible, and it comes with 26 different editing filters.

Photobook options

The Fujifilm X-T3 digital camera is an attempt to level the playing field in terms of access to professional-grade photography.

A built-in photobook helper can take pictures of individuals and places for you.

Menus for camera

With the Fujifilm X-T3, you may change several camera settings through the System Menu. This menu is the main interface for the camera, allowing the user to change settings including shutter speed, aperture, ISO, and white balance.

A simple click on the System Menu will take you to all of these settings.

Focus mode and image quality are only two of the many settings that may be modified.

System menus may also be found on several other Fujifilm cameras, including the X-T2, X-H1, and X-T20.

Settings for user usage

Fujifilm's X-T3 is a small camera with a big sensor. It's not just one of the greatest cameras you can buy, but also one of the cheapest ones.

The Fujifilm X-T3 is a tiny camera with a 23.5 x 15.6mm CMOS sensor, making it capable of producing high-quality photographs with low levels of noise, even in low-light settings.

The camera offers an ISO range of 100 to 12800 (expandable to 25600) and a rapid shutter speed of 1/4000, making it ideal for photographing fast-moving subjects or in low light without the blur often associated with longer shutter speeds.

Built-in Wi-Fi and Bluetooth, GPS, 4K video recording, electronic viewfinder (EVF), 3" LCD screen with touch screen capability; these are just a few of the many convenient features included on the Fujifilm X-T3.

Settings for sound

The Fujifilm X-T3 is a high-quality camera that records audio very well.

It is capable of recording high-quality sound in every shooting environment and capturing clear audio even in low light.

Power Source

The Fujifilm X-battery S10's life is among the best of any digital camera currently available.

Depending on the circumstances, it may record for up to 100 hours without recharging in a single shot.

Screen menu

The Fujifilm X-T3 is a pocketable tiny camera. It has a tiny 20-megapixel sensor and a wide-angle 24 mm lens.

With its compact sensor and wide-angle lens, the Fujifilm X-T3 is an ideal choice for beginners.

As a result, you can take high-quality images even when lighting conditions are less than ideal.

Built-in Wi-Fi and Bluetooth allow you to wirelessly transfer photos and videos to a mobile device.

Menu for dials

You may quickly and simply adjust the X-settings S10's using the device's convenient button/dial interface.

When shooting in manual mode, which may be difficult to control without glancing at the camera screen, this option comes in very handy.

The X-T3 includes a lever on the rear that can be slid up and down for use in manual mode. Utilizing this lever allows for instantaneous adjustments to both ISO and exposure.

Management of power

Simply said, the Fujifilm X-T3 is a basic DSLR camera with built-in power management technology that prevents the camera from draining the battery until it is in use.

This camera has a number of manual options and an electronic viewfinder, making it ideal for more advanced photographers.

The Fujifilm X-T3 may be customized with a variety of lenses.

The camera's electronic shutter makes it possible to take pictures without making a sound.

If you're just starting out in photography but want to take photos of a professional quality, but you're on a small budget, the Fujifilm X-T3 is a great choice.

Saving data

An SD card slot is included in the Fujifilm X-T3 digital camera. It can take images and videos of excellent quality.

The camera has both Wi-Fi and Bluetooth connectivity right out of the box.

However, you can't make the SD card the default destination for media files like photos and movies. Here are some potential answers to this predicament:

1) Connect the camera to the computer via USB connection and download the photos.

2) Use a memory card reader to move images from your computer to the camera, and save images and movies in the camera's internal memory until you can move them.

Settings for connections

The Fujifilm X-T3 is a compact camera that can be used wirelessly and fits easily into a pocket. Connecting the camera to a mobile device facilitates remote control.

The camera's connection setting is one of its unique selling points if you're a photographer who wants to take better images without being distracted by their equipment.

The Fujifilm X-T3 has a straightforward UI thanks to its three control dials and large buttons.

This camera has an eye-level viewfinder and LCD screen.

The X-T3 is a compact camera that, thanks to its image stabilizer, can take clear pictures even in low light.

It also has Wi-Fi connectivity, allowing you to submit photographs to social media platforms like Instagram and Facebook from the convenience of your mobile device.

Networks

Discover in this part why the Fujifilm X-T3 is a very competent camera with excellent connection.

Connectivity-wise, the Fujifilm X-T3 digital camera excels.

To connect and take control, all you need is a mobile device, tablet, or computer.

The camera app on your phone may be used for remote control.

With the built-in Wi-Fi, you may easily transfer your photos and videos to other devices.

Using HDMI

The ability to connect an HDMI monitor or TV to a camera is enabled by the inclusion of an HDMI Output. It's not present on all cameras and often resides either at the camera's top or its rear.

The X-T3 from Fujifilm can send data to an HDMI monitor.

This allows you to connect your camera to any device with an HDMI output without the need for additional connections or adapters.

More and more people are opting for HDMI Outputs because they simplify the often-confusing process of connecting devices with a variety of various types of connections.

USB connections

Here, you'll get the lowdown on the many USB choices for using your Fujifilm X-T3.

Connecting a Fujifilm camera to a computer is as easy as plugging in a USB cable. In particular, you should proceed as follows.

The first thing you need to do is connect the camera to a computer using the USB cable.

2) Open the Camera Wizard in Windows or Mac OS X.

Step 3: Follow the wizard's directions to set up your camera and install the software.

Connect your camera to your computer again via USB and proceed with steps 1 and 2 from above to upload your photographs.

5. Disconnect the device from the computer by severing the USB cable.

Fujifilm X-T3 Computer Connection

To begin, a Fujifilm X-T3 camera must be connected to a computer through a USB cable.

To establish a connection between the cameras, the cable must be inserted into both ports simultaneously (one end plugs into your camera, the other into the computer).

Next, in either Windows or Mac OS X, start
the Camera Wizard and follow the on-screen
instructions to get the necessary software.
Before to connecting the Fujifilm NP-
W126S battery to a computer, make sure it
has reached 100% charge.

When the program is loaded, the camera
may be connected to a computer through
USB to transfer images.

Lastly, disconnect each device from the
Computer by unplugging the USB cable
from its end at the same time.

Apparatus

The Fujifilm X-T3 is a compact and lightweight point and shoot that captures images at a high quality.

It has a 20.3 megapixel sensor and a 24 millimeter wide-angle lens, allowing you to snap photographs that are both attractive and suitable for sharing or using as wallpaper.

If you're a photographer who prioritizes mobility and affordability when on the road, the Fujifilm X-T3 is a fantastic choice.

Because to its high-resolution 20.3-megapixel sensor, the Fujifilm X-T3 is capable of producing images with little noise.

The 24mm wide-angle lens allows you to capture beautiful scenery or cityscapes without resorting to extreme close-ups or far-reaching zooms.

With its user-friendly design, wide-angle lens, and 20.3-megapixel sensor, the

Fujifilm X-T3 makes for a great entry-level camera.

Using Lenses

The Fujifilm X-T3 is an APS-C camera with 24.2 megapixels and a wide range of user-friendly features that make it accessible to photographers of all skill levels.

The Fujifilm X-T3 is the best budget option with all the bells and whistles for photographers.

The APS-C sensor's 24.2 megapixel resolution and the camera's ability to capture 10 frames per second continuously are noteworthy.

Getting the appropriate lens for your camera is essential.

The Fujifilm X-T3 comes with a couple of lenses: a 45mm f/4.0 and an 18mm f/2.0 dual-focus prime (with manual focus).

The autofocus lens is great for low-light conditions, while the manual-focus lens allows you greater creative control over the composition of your shots.

The Fujifilm X-3D T3's panoramic mode allows you to capture breathtaking, shareable panoramas.

Framing your shots is a breeze thanks to the camera's 460K-dot OLED viewfinder and 1.44 million-dot built-in electronic viewfinder.

Capturing gorgeous, high-detail images is a breeze with the Fujifilm X-12MP S10, thanks to its APS-C sensor.

The Fujifilm X-T3 is capable of recording video in 4K Ultra High Definition (UHD) resolution at 60 frames per second with an ISO range of 100–510000.

With the filter and Focus Bracketing functions of the Fujifilm X-Low-Pass S10, you can take photos that are both clear and visually pleasing.

The X-T3 is a versatile camera that can take high-quality pictures in a variety of lighting circumstances.

EFU

For photographers, the external flash unit is the single most useful piece of equipment.

They inspire photographers to be creative by letting them experiment with different lighting techniques.

The Fujifilm X-T3 is highly recognized as an external flash unit because to its user-friendliness and practical characteristics.

It may be used as a wireless trigger and has an adjustable light intensity as well as a wireless receiver.

Radio-Activated Shock

Among the many amazing features of the Fuji X-ability S10 is its capacity to act as a radio trigger.

As a result, photographers may use their other Fujifilm cameras without being restricted by their own equipment, allowing them to try out different compositions and lighting arrangements.

Photographers may utilize the smartphone's camera as a remote shutter and the X-three S10's levels of lighting to take gorgeous portraits and landscape shots.

Conclusion

The Fujifilm X-T3 is a high-quality camera that has many useful functions and can be used effectively by both amateurs and professionals.

It can autofocus quickly and precisely, record 4K video, and has an intervalometer.

According to this book's findings, the Fujifilm X-T3 is a fantastic camera that offers plenty of bang for the buck.

Made in the USA
Middletown, DE
24 April 2023